YOUR KNOWLEDGE HAS

Bibliographic information published by the German National Library:

The German National Library lists this publication in the National Bibliography; detailed bibliographic data are available on the Internet at http://dnb.dnb.de .

Imprint:

Copyright © 2018 GRIN Verlag
Print and binding: Books on Demand GmbH, Norderstedt Germany
ISBN: 9783668792500

This book at GRIN:

https://www.grin.com/document/440197

Saranya Karunamurthi

Braille Learning System using Raspberry Pi

GRIN Verlag

GRIN - Your knowledge has value

Since its foundation in 1998, GRIN has specialized in publishing academic texts by students, college teachers and other academics as e-book and printed book. The website www.grin.com is an ideal platform for presenting term papers, final papers, scientific essays, dissertations and specialist books.

Visit us on the internet:

http://www.grin.com/

http://www.facebook.com/grincom

http://www.twitter.com/grin_com

ABSTRACT

Braille literacy is crucial for blind individuals, as it enables life-long learning and is key to employment and independency. One way to promote Braille literacy is to make existing Braille reading devices more accessible, affordable, and user friendly. Commercially available Braille reading devices in this regard need various improvements. Most of these devices cost thousands of dollars, mainly because they rely on multiple piezoelectric actuators in order to create the Braille letters. Other issues include high voltage actuation and reduced portability. In order to solve these issues and improve existing Braille displays, various actuation methods have been widely investigated.285 million people are estimated to be visually impaired worldwide, 39 million are blind and 246 have low vision. India has the largest population of blind people in the world. That's over 12 million people. Even the central and state government provide job opportunities for blind and visually impaired people. But only few vacancies are filled due to lack of education. Lack of education for blind people is due to the lack of qualified teachers to teach them Braille. Learning the Braille script is not an easy task for Visually Impaired students. Visually Impaired students have to memorize various patterns of keys of Braille matrix assigned for different letters/words/symbols in Braille script to read and write effectively. This project helps in simplifying the process of learning of braille instructions. The integration of physical activity and hearing can facilitate easy learning of Braille Script. In this project we develop a kit which helps the blind and visually impaired people to learn braille even in the absence of teachers.

Inhalt

LIST OF FIGURES

5

LIST OF TABLES

CHAPTER 1

INTRODUCTION

1.1 BLOCK DIAGRAM

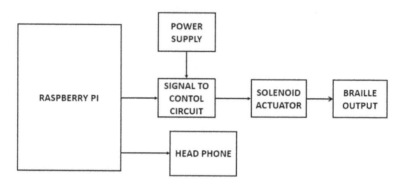

Fig 1.1 Block diagram of the Proposed system

The raspberry pi is coded in python. Accordingly the raspberry pi produces the output and its output signal is sent to the solenoid actuator which acts as the braille display. The corresponding audio output is taken from the raspberry pi and it is heard in the headphone.

1.2 ABOUT BARILLE

Braille is named after its creator, Frenchman Louis Braille, who lost his eyesight due to a childhood accident. Braille was invented in 1824, the simple dot system is still considered by many people as the best reading system for blind people. It is a brilliant tool for communicating and participating in society. Braille is a tactile writing system used by people who are blind or visually impaired. It is traditionally written with embossed paper. Braille is a system of raised dots that can be read with the fingers by people who are blind or who havelow vision. Braille is not a language. It is a code by which all languages can be written and read. Through the use of Braille, people who are blind are able to review and study the written other word. It provides a passport for literacy and gives an individual the ability to become familiar with spelling, punctuation, paragraphing and formatting considerations.

1.3 BRAILLE STRUCTURE

Braille symbols are formed within units of space known as Braille cells. A full Braille cell consists of six raised dots arranged in two parallel vertical columns of three dots (like the number 6 on a dice). The dot positions are identified by numbers one through to six. 63 combinations are possible using one or more of these six dots. Cells can be used to represent a letter of the alphabet, number, punctuation, part of a word or even a whole word.

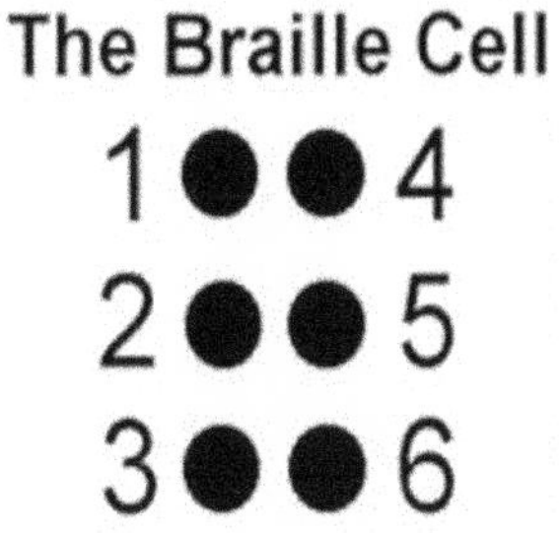

The Braille Cell

1 ● ● 4

2 ● ● 5

3 ● ● 6

Fig 1.2 Arrangement of Braille cell

1.4 IMPORTANCE OF BRAILLE

The expectations for students to learn to read and to write is no different for a sighted or blind person; it is fundamental skill that everyone should develop in order to excel in their life. There is no substitute for the ability to read. For blind people, braille is an essential tool that helps in the process of becoming literate. Tape recorders and synthesized speech are useful tools, but they are inadequate substitutes for reading and writing. Braille plays an important role in education of blind and visually impaired people. Several studies show that braille literacy is directly related with academic achievement and employment among the blind and visually impaired. There are many government job available for blind and visually impaired people. They must be educated to get job from the government organisation. For that they must learn braille which is an important thing. When blind children learn Brailje, they learn grammar, spelling, punctuation and sentence structure skills that they do not learn using text-to-speech technology. Braille also lets blind people read charts and graphs that are almost impossible to convey using text-to-speech. Even if they also use low-vision aids or text-to-speech technology, Braille lets them label items clearly, read public signs at airports or bathrooms.

8

Fig 1.3 Braille in Rubik's cube

1.5 BRAILLE IN DAY TO DAY LIFE

A person who reads braille can roam independently among the community as braille increasingly is showing up everywhere. With the Convention on the Rights of Persons with Disabilities implemented in 2008, since then over 150 countries have signed the ratification. The declaration has introduced the recognition of persons with disabilities on an equal basis with others worldwide. Additionally, society is beginning to recognize the need to present equality for all citizens within its communities. Therefore, public spaces continue to present tools and aids for all disabilities. Braille is used in everyday communication and as a literate blind or visually impaired individual, independence is given. Blind individuals deserve this chance at equality, and this is something that braille provides. Speech feedback or other digital tools are no compensation for braille. Listening alone is not enough. To read without braille, a person who is blind is entirely dependent on computers with voice synthesizers or audio recording, neither of which is useful in every circumstance. Braille brings sight to the visually impaired and blind.

Fig 1.4 Braille in ATM keypad

1.6 BRAILLE SYMBOLS

The basic braille alphabet, braille numbers, braille punctuation and special symbols characters are constructed from six dots. These braille dots are positioned like the figure six on a die, in a grid of two parallel vertical lines of three dots each. From the six dots that make up the basic grid, 64 different configurations can be created.

1.6.1 THE BRAILLE ALPHABET

Table no 1.6.1 Braille cell arrangement for alphabets

CHARACTER	BRAILLE	BRAILLE DOTS POSITION
A	•	1
B	⁚	1,2
C	••	1,4
D	•⁚	1,4,5
E	•.	1,5
F	⁚•	1,4,2
G	⁚⁚	1,4,2,5

10

H	⠓	1,2,5
I	⠊	4,2
J	⠚	4,2,5
K	⠅	1,3
L	⠇	1,2,3
M	⠍	1,4,3
N	⠝	1,4,5,3
O	⠕	1,5,3
P	⠏	1,2,3,4
Q	⠟	1,2,3,4,5
R	⠗	1,2,3,5
S	⠎	2,3,4
T	⠞	2,3,4,5
V	⠧	1,2,6
U	⠥	1,3,6
W	⠺	2,4,5,6
X	⠭	1,4,3,6
Y	⠽	1,3,4,5,6
Z	⠵	1,3,5,6

1.6.2 THE BRAILLE NUMBERS

Braille numbers are formed by placing a braille number sign before braille letters. They are internationally accepted braille system.

Table no 1.6.2 Braille cell arrangement for numbers

NUMBERS	BRAILLE	POSITION OF BRAILLE DOTS
1		3,4,5,6, 1
2		3,4,5,6, 1,2
3		3,4,5,6, 1,4
4		3,4,5,6, 1,4,5
5		3,4,5,6,1,5
6		3,4,5,6, 1,2,4
7		3,4,5,6, 1,2,5
8		3,4,5,6, 1,2,5
9		3,4,5,6,2,4

1.6.3 THE BRAILLE PUNCTUATION

These are the common punctuation used in our day to day life. These braille combinations for punctuations are accepted internationally.

Table no 1.6.3 Braille cell arrangement for punctuations

PUNCTUATION	BRAILLE	POSITION OF BARILLE DOTS
,	•	2
;	:	2,3
:	••	2,5
.	•:	2,5,6
!	:•	3,2,5
?	:.	2,3,6
"	• ::	3,5,6
'	• .:	3,3,5,6
-	• .:	3,6

CHAPTER 2

LITERATURE SURVEY

Saurabh Bisht,Sandeep Reddy Goluguri,Rajat Maheshwari,Akhilesh Kumar and P.Sathya[1]

The aim of this project is to create a refreshable electronic braille display using Raspberry Pi 2 and Arduino Mega2560 board. This aims at creating a refreshable braille display that is capable of converting Normal text file as well as printed image files to braille. To achieve the functionality of converting image files to text we have used tesseract-ocr engine of Google which the best available OCR engine available right now and is highly accurate. The Raspberry pi does the image processing and the Arduino drives the display. The display is made of 6 servo motors, which can be controlled by the PWM. The servo motor acts as an actuator in this project.

Prachi Rajarapollu, Stavan Kodolikar, Dhananjay Laghate, Amarsinh Khavale[2]

Blind people are an integral part of the society. However, their disabilities have made them to have less access to computers, the Internet, and high quality educational software than the people with clear vision. Consequently, they have not been able to improve on their own knowledge, and have significant influence and impact on the economic, commercial, and educational ventures in the society. One way to narrow this widening gap and see a reversal of this trend is to develop a system, within their economic reach, and which will empower them to communicate freely and widely using the Internet or any other information infrastructure. Over time, the Braille system has been used by the visually impaired for communication and contact with the outside world. This paper presents the implementation of Braille to Text/Speech Converter on FPGA Spartan3 kit. The actual Braille language is converted into English language in normal domain. The input is given through braille keypad which consists of different combinations of cells. This input goes to the FPGA Spartan3 Kit. According to the combinations given, FPGA converts the input into corresponding english text through the decoding logic in VHDL language. After decoding, the corresponding alphabet is converted to speech through algorithm. Also it is displayed on the LCD by interfacing the LCD to the Spartan3 kit.

Aisha Mousa, Hazem Hiary, Raja Alomari, and Loai Alnemer [3]

Firsthand Experiences with a Balanced Approach to Literacy, new from the American Foundation for the Blind, begins with a discussion of the whole language and traditional approaches to teaching reading and writing. Author Anna M. Swenson is a Braille teacher who favors the whole language philosophy, but who has taught in various settings, and includes ideas and techniques that will work with the traditional approach as well. The books focus is on creating an atmosphere that promotes literacy, no matter what the teaching approach.According to Swenson, Braille teachers, who are the book primary intended audience, are not only teaching the Braille code, but are also teaching reading and writing. Swenson encourages Braille teachers to keep up to date with current approaches to teaching language arts, to get samples of sighted students work in order to understand the level of classroom expectation, and to consult with the reading specialist or learning disabilities specialist in the school if the blind student seems to be having difficulty learning to read.Swenson makes detailed suggestions regarding working out the technical aspects of teaching Braille to a student in the mainstream, especially in a whole language classroom. In the section A Morning in the Mainstream, the reader can get a vivid view of how the Braille teacher can work alongside the classroom teacher to ensure a solid foundation for the blind student.

Vineeth Kartha; Dheeraj S. Nair; S. Sreekant; P. Pranoy; P. Jayaprakash[4]

Braille is a textile document which is most common means of reading for visually challenged people. There is a significant need to preserve old Braille documents and literatures. Hard-copy Braille is heavy, bulky, which takes considerable space to store and maintain. The materials that are available only in hard copy bear a threat of losing them. The Automation of Braille reading and digitization process will help in preserving, distributing Braille over network and reproducing hard copy only on demands. The need for digitizing Braille has lead many researchers in improvising and implementation of optical Braille recognition systems. A survey and comparative study on various algorithms and techniques proposed and presented by researchers in improvising Optical Braille Recognition system to effectively/efficiently translate Braille to text has been presented in this paper.

CHAPTER 3

HARDWARE DESCRIPTION

3.1 Raspberry Pi 3 Model B

Fig 3.1 Raspberry pi 3 processor

The Raspberry Pi 3 is the third generation Raspberry Pi. It is a series of small single board computers developed in the United Kingdom by the Raspberry foundation to promote the teaching of basic computer science in schools and in developing countries. It builds upon the features of its predecessors with a new, faster processor on board to increase its speed. The Raspberry Pi hardware has evolved through several versions that feature variations in memory capacity and peripheral-device support. It has an identical form factor to the previous Pi 2 and has complete compatibility with Pi 1 and 2.

16

3.1.1 PINDIAGRAM

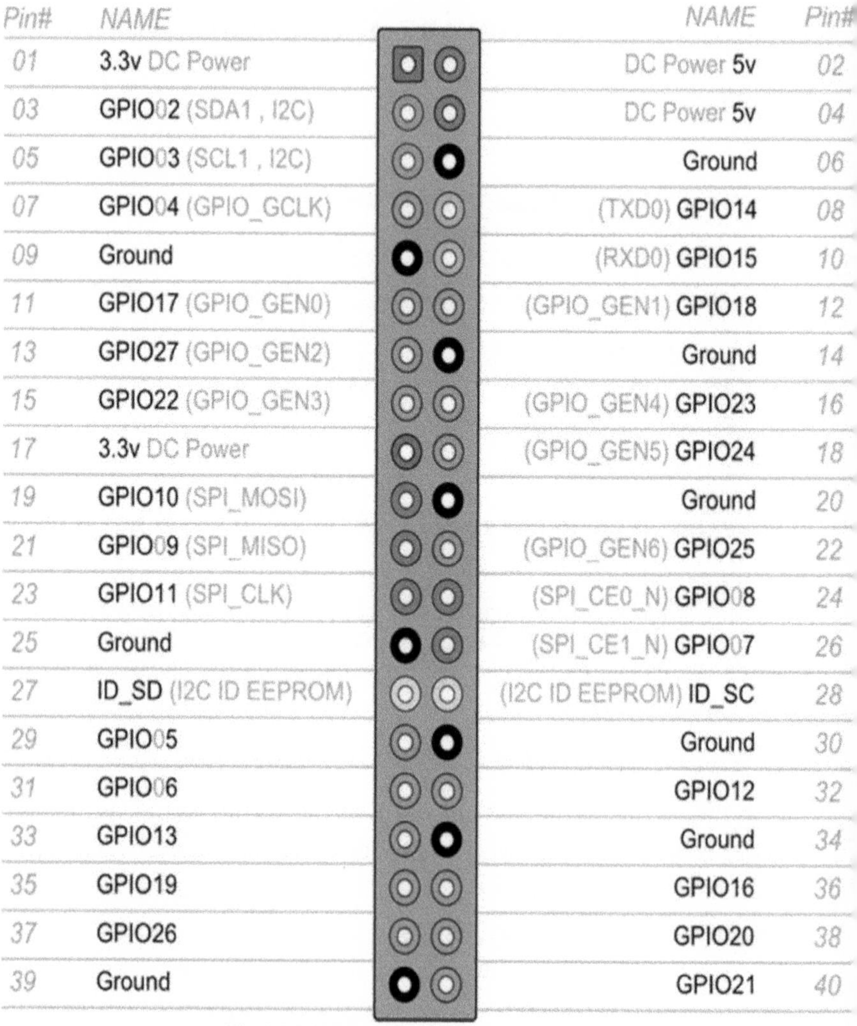

Pin#	NAME			NAME	Pin#
01	3.3v DC Power	◨ ◯		DC Power 5v	02
03	GPIO02 (SDA1 , I2C)	◯ ◯		DC Power 5v	04
05	GPIO03 (SCL1 , I2C)	◯ ●		Ground	06
07	GPIO04 (GPIO_GCLK)	◯ ◯		(TXD0) GPIO14	08
09	Ground	● ◯		(RXD0) GPIO15	10
11	GPIO17 (GPIO_GEN0)	◯ ◯		(GPIO_GEN1) GPIO18	12
13	GPIO27 (GPIO_GEN2)	◯ ●		Ground	14
15	GPIO22 (GPIO_GEN3)	◯ ◯		(GPIO_GEN4) GPIO23	16
17	3.3v DC Power	◯ ◯		(GPIO_GEN5) GPIO24	18
19	GPIO10 (SPI_MOSI)	◯ ●		Ground	20
21	GPIO09 (SPI_MISO)	◯ ◯		(GPIO_GEN6) GPIO25	22
23	GPIO11 (SPI_CLK)	◯ ◯		(SPI_CE0_N) GPIO08	24
25	Ground	● ◯		(SPI_CE1_N) GPIO07	26
27	ID_SD (I2C ID EEPROM)	◯ ◯		(I2C ID EEPROM) ID_SC	28
29	GPIO05	◯ ●		Ground	30
31	GPIO06	◯ ◯		GPIO12	32
33	GPIO13	◯ ●		Ground	34
35	GPIO19	◯ ◯		GPIO16	36
37	GPIO26	◯ ◯		GPIO20	38
39	Ground	● ◯		GPIO21	40

Fig 3.2 Pindiagram of Raspberry pi

The Raspberry Pi may be operated with any generic The Broadcom BCM2835 SoC used in the first generation Raspberry Pi is somewhat equivalent to the chip used in first generation Smartphone's (its CPU is an older ARMv6 architecture), which includes a 700 MHz ARM1176JZF-S processor, Video Core IV graphics processing unit (GPU), and RAM.

It has a level 1 (L1) cache of 16 KB and a level 2 (L2) caches of 128 KB. The level 2 cache is used primarily by the GPU. The SoC is stacked underneath the RAM chip, so only its edge is visible.The current Model B boardsincorporate four USB ports for connecting peripherals like computer, keyboard and mouse.It may also be used with USB storage, USB to MIDI converters, and virtually any other device/component with USB capabilities.Other peripherals can be attached through the various pins and connectors on the surface of the Raspberry Pi. Raspberry pi can be used with varoius operating system. There are linux based and non linux based oprating system. The mainly used operatind system is Raspian.

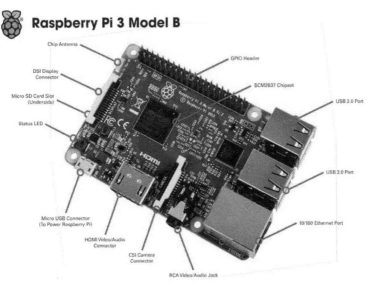

Fig 3.3 Technical parts of Raspberry pi 3

3.2 SOLENOID ACTUATOR

Fig 3.4 Solenoid

It is an actuator that electrical signal into producing a linear

actuator electromagnetic converts an a magnetic field motion. It works on

18

the principle of electromagnetic relay. It consists of an electrical coil wound around a magnetic actuator or plunger that is free to move or slide IN and OUT of the coils body. Solenoids can be used to electrically open doors and latches, open or close valves, and to actuate electrical switches just by energizing its coil. It can also be designed for proportional motion controls were the plunger position is to the power input. Linear solenoids are useful in many applications that require an open or closed type motion such as electronically activated door locks, pneumatic or robotics.

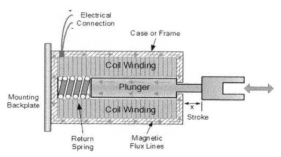

Fig 3.5 Internal View of Solenoid Actuator

When an electrical current is passed through the coils windings, it behaves like an electromagnet and the plunger which is located inside the coil, is attracted towards the center of the coil by the magnetic flux setup within the coils body, which in turn compresses a small spring attached to one end of the plunger. The force and speed of the plungers movement is determined by the strength of the magnetic flux generated within the coil. When the supply current is turned OFF (de energized) the electromagnetic field generated previously by the coil collapses and the energy stored in the compressed spring forces the plunger back out to its original rest position. This back and forth movement of the plunger is known as the solenoids Stroke. Linear solenoids are available in two basic configurations called a Pull type as its pulls the connected load towards itself when energized, and the Push type that act in the opposite direction pushing it away from itself when energized. Both types are generally constructed the same with the difference in location of the spring and design of the plunger.

3.3 TRANSISTOR TIP122

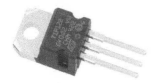

Fig 3.6 Transistor TIP122

Table no 3.3.1 Characteristics of Transistor TIP122

Characteristics	Range
Transistor Type	NPN
Maximum Continuous Collector Current	8 A
Maximum Collector Emitter Voltage	100 V
Maximum Emitter Base Voltage	5 V
Pin Count	3
Transistor Configuration	Single
Number of Elements per Chip	1
Minimum DC Current Gain	1000
Maximum Collector Base Voltage	100 V
Maximum Collector Emitter Voltage	4V
Maximum Collector Cut-off Current	0.2Ma
Minimum Operating Temperature	-65 °C
Height	9.15mm
Maximum Operating Temperature	+150 °C

Length	10.4mm
Width	4.6mm
Dimensions	10.4 x 4.6 x 9.15mm

The TIP122 is an NPN Transistor. In that format, we know that the base requires a voltage that is more positive than the emitter in order to allow current to flow from the emitter to the collector.

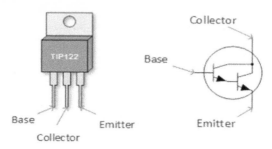

Fig 3.7 Symbol of Transistor

The transistor allows as much as 5 amps of current going from the emitter through to the collector and 120 mA from the emitter through to the base. It has 100-volt difference from the collector to the emitter and as much as a 100 volt between the collector and base.In electronics, the Darlington transistor (often called a Darlington pair) is a compound structure consisting of two bipolar transistors (either integrated or separated devices) connected in such a way that the current amplified by the first transistor is amplified further by the second one. This configuration gives a much higher current gain than each transistor taken separately and, in the case of integrated devices, can take less space than two individual transistors because they can use a shared collector. Integrated Darlington pairs come packaged singly in transistor-like packages or as an array of devices (usually eight) in an integrated circuit. A Darlington pair can be sensitive enough to respond to the current passed by skin contact even at safe voltages. Thus it can form the input stage of a touch-sensitive switch. Darlington transistors can be used in circuits involving motors, relays, or other current-hungry components connected to computers.

3.4 DIODE 1N4007

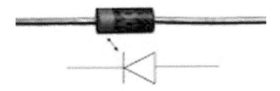

Fig 3.8 Diode 1N4007

A rectifier diode is used as a one-way check valve. Since these diodes only allow electrical current to flow in one direction, they are used to convert AC power into DC power. When constructing a rectifier, it is important to choose the correct diode for the job; otherwise, the circuit may become damaged. A 1N4007 diode is electrically compatible with other rectifier diodes, and can be used as a replacement for any diode in the 1N400x family. A diode allows electrical current to flow in one direction from the anode to the cathode. Therefore, the voltage at the anode must be higher than at the cathode for a diode to conduct electrical current. When the voltage at the cathode is greater than the anode voltage, the diode will not conduct electrical current. In practice, the diode conducts a small current under these circumstances. If the voltage differential becomes great enough, the current across the diode will increase and the diode will break down. Some diodes such as the 1N400 will break down at 50 volts or less. The 1N4007, can sustain a peak repetitive reverse voltage of 1000 volts.

When the voltage at the anode is higher than the cathode voltage, the diode is said to be "forward-biased," since the electrical current is "moving forward." The maximum amount of current that the diode can consistently conduct in a forward-biased state is 1 ampere. The maximum that the diode can conduct at once is 30 amperes. However; if the diode is required to conduct that much current at once, the diode will fail in approximately 8.3 milliseconds. When the maximum allowable consistent current amount is flowing through the diode, the voltage differential between the anode and the cathode is 1.1 volts. Under these conditions, a 1N4007 diode will dissipate 3 watts of power (about half of which is waste heat).

3.5 RESISTOR 1K

Fig 3.9 Resistor 1K

A resistor is a passive two-terminal electrical component that implements electrical resistance as a circuit element. In electronic circuits, resistors are used to reduce current flow, adjust signal levels, to divide voltages, bias active elements, and terminate transmission lines, among other uses. High-power resistors that can dissipate many watts of electrical power as heat may be used as part of motor controls, in power distribution systems, or as test loads for generators. Fixed resistors have resistances that only change slightly with temperature, time or operating voltage. Resistors are common elements of electrical networks and electronic circuits and are ubiquitous in electronic equipment. Practical resistors as discrete components can be composed of various compounds and forms. Resistors are also implemented within integrated circuits. The electrical function of a resistor is specified by its resistance: common commercial resistors are manufactured over a range of more than nine orders of magnitude. The nominal value of the resistance falls within the manufacturing tolerance, indicated on the component.

CHAPTER 4

POWER SUPPLY DESCRPITION

4.1 7812 REGULATOR

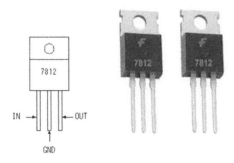

Fig 4.1 Pin diagram of 7812 regulator

The L7812 is a low voltage regulator able to provide up to 1A of Output Current at a fixed 12V output voltage.

Output Voltage:12V

Output Current:1A

Package / Case:TO-220

Input Voltage Max:35V78xx (sometimes L78xx, LM78xx, MC78xx...) is a family of self-contained fixed linear voltage regulator integrated circuits. The 78xx family is commonly used in electronic circuits. For ICs within the 78xx family, the xx is replaced with two digits, indicating the output voltage (for example, the 7805 has a 5-volt output, while the 7812 produces 12 volts). The 78xx line are positive voltage regulators: they produce a voltage that is positive relative to a common ground. There is a related line of 79xx devices which are complementary negative voltage regulators. 78xx and 79xx ICs can be used in combination to provide positive and negative supply voltages in the same circuit requiring a regulated power supply due to their ease-of-use and low cost.These devices support an input voltage anywhere from around 2.5 volts over the intended output voltage up to a maximum of 35 to 40 volts depending on the model, and typically provide 1 or 1.5 amperes of current (though smaller or larger packages may have a lower or higher current rating).

4.2 CAPACITOR 470uF

Fig 4.2 Capacitor(470uF)

A capacitor is a passive two-terminal electrical component that stores electrical energy in an electric field. The effect of a capacitor is known as capacitance.Most capacitors contain at least two electrical conductors often in the form of metallic plates or surfaces separated by a dielectric medium. A conductor may be a foil, thin film, sintered bead of metal, or an electrolyte. The non-conducting dielectric acts to increase the capacitor's charge capacity. Materials commonly used as dielectrics include glass, ceramic, plastic film, paper, mica, and oxide layers. Capacitors are widely used as parts of electrical circuits in many common electrical devices. Unlike a resistor, an ideal capacitor does not dissipate energy.Capacitors are widely used in electronic circuits for blocking direct current while allowing alternating current to pass. In analog filter networks, they smooth the output of power supplies. In resonant circuits they tune radios to particular frequencies. In electric power transmission systems, they stabilize voltage and power flow. The property of energy storage in capacitors was exploited as dynamic memory in early digital computers.High temperature electrolytic for use in switch mode power supplies found in TVs, VCRs, etc. 105° C operating temperature satisfies characteristic W of JIS-C-5141 standards. ± 20% tolerance.

4.3 TRANSFORMER

Atransformer is an electrical device that transfers electrical energy between two or more circuits through electromagnetic induction. A varying current in one coil of the transformer produces a varying magnetic field, which in turn induces a voltage in a second coil. Power can be transferred between the two coils through the magnetic field, without a metallic connection between the two circuits. Faraday's law of induction discovered in 1831 described this effect. Transformers are used to increase or decrease the alternating voltages in electric power applications.

Fig 4.3 Transformer(220V AC to 24V AC)

Since the invention of the first constant-potential transformer in 1885, transformers have become essential for the transmission, distribution, and utilization of alternating current electrical energy. A wide range of transformer designs is encountered in electronic and electric power applications. Transformers range in size from RF transformers less than a cubic centimeter in volume to units interconnecting the power grid weighing hundreds of tons.Since the high voltages carried in the wires are significantly greater than what is needed in-home, transformers are also used extensively in electronic products to decrease (or step-down) the supply voltage to a level suitable for the low voltage circuits they contain. The transformer also electrically isolates the end user from contact with the supply voltage. Transformers are used to increase (or step-up) voltage before transmitting electrical energy over long distances through wires. Wires have resistance which loses energy through joule heating at a rate corresponding to square of the current. By transforming power to a higher voltage transformers enable economical transmission of power and distribution. Consequently, transformers have shaped the electricity supply industry, permitting generation to be located remotely from points of demand.

4.4 REGULATED POWER SUPPLY WORKING

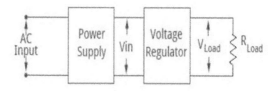

Fig 4.4 Power supply Block Diagram

A transformer operates on the principals of "electromagnetic induction", in the form of Mutual Induction.Mutual induction is the process by which a coil of wire magnetically induces a voltage into another coil located in close proximity to it.transformers are capable of either increasing or decreasing the voltage and current levels of their supply, without modifying its frequency, or the amount of electrical power being transferred from one winding to another via the magnetic circuit.A transformer that increases voltage from primary to secondary (more secondary winding turns than primary winding turns) is called a step-up transformer. Conversely, a transformer designed to do just the opposite is called a step-down transformer.Full wave bridge rectifier type of single phase rectifier uses four individual rectifying diodes connected in a closed loop "bridge" configuration to produce the desired output. The main advantage of this bridge circuit is that it does not require a special center tapped transformer, thereby reducing its size and cost. The single secondary winding is connected to one side of the diode bridge network and the load to the other side.

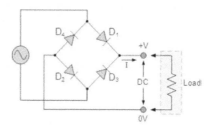

Fig 4.5 Full wave bridge rectifier

The four diodes labelled D1 to D4 are arranged in "series pairs" with only two diodes conducting current during each half cycle. During the positive half cycle of the supply,diodesD1 and D2 conduct in series while diodes D3 and D4 are reverse biased.

4.4.1 THE POSITIVE HALF-CYCLE

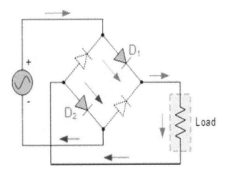

Fig 4.6 Positive half cycle of full wave bridge rectifier

During the Positive half cycle of the input AC waveform diodes D1 and D2 are forward biased and D3 and D4 are reverse biased. When the voltage, more than the threshold level of the diodes D1 and D2, starts conducting – the load current starts flowing through it.

4.4.2 THE NEGATIVE HALF-CYCLE

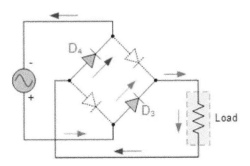

Fig 4.7 Negative half cycle of full wave bridge rectifier

During the negative half cycle of the input AC waveform, the diodes D3 and D4 are forward biased, and D1 and D2 are reverse biased. Load current starts flowing through the D3 and D4 diodes.

The single phase half-wave rectifier produces an output wave every half cycle and that it was not practical to use this type of circuit to produce a steady DC supply. The full-wave bridge rectifier, gives us a greater mean DC value (0.637 Vmax) with less superimposed ripple while the output waveform is twice that of the frequency of the input supply frequency. The average DC output of the rectifier can be improved while at the same time reducing the AC variation of the rectified output by using smoothing capacitors to filter the output waveform. Smoothing or reservoir capacitors connected in parallel with the load across the output of the full wave bridge rectifier circuit increases the average DC output level even higher as the capacitor acts like a storage device. 7812 input voltage range is 14V to 35V. Exceeding the voltage range may damage the IC. The output from the capacitor is given to 7812 to regulator to get constant output voltage of about 12v and 1A. Voltage regulators stabilize the output voltage against variations in input voltage. Ripple is equivalent to a periodic variation in the input voltage. Thus, a voltage regulator attenuates the ripple that comes in with the unregulated input voltage. Since a voltage regulator uses negative feedback, the distortion is reduced by the same factor as the gain.

CHAPTER 5

5.1 CIRCUIT DIAGRAM OF CONTROL UNIT

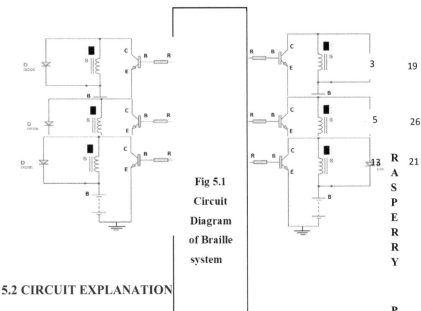

Fig 5.1
Circuit
Diagram
of Braille
system

3   19

5   26

13   R   21
     A
     S
     P
     E
     R
     R
     Y

     P
     I

5.2 CIRCUIT EXPLANATION

Raspberry pi is a processor and it produces the control signals. Raspberry pi GP in produces an output of 3.3volts and 50mA approximately. Here we have used six GPI ns of Raspberry pi say, pin no 3,5,13,19,21,26, pin no 6 is used as ground. Its output ot sufficient to energize the solenoid actuator. Hence we use external power supply.To energize the solenoid actuator here we have designed a power supply. Raspberry pi entirely coded in python to produce the desired output.In Raspberry pi if the pin becomes high, it sends the control signal to the base of the transistor. The base of the transistor is triggered and the circuit is closed. Once the circuit is closed the solenoid actuator draws power from the power supply and the actuator is energized.The transistor (TIP122) is used as switch in the circuit.For example, we take letter A, for this the solenoid actuator connected to the pin no 5,13,19,21,26 will energize. This will form the braille combination of letter A. The diode(1N4007) is connected across the solenoid actuator to avoid the voltage spike from damaging the device.

5.2.1 OPEN CIRCUIT CONDITION OF SOLENOID ACTUATOR

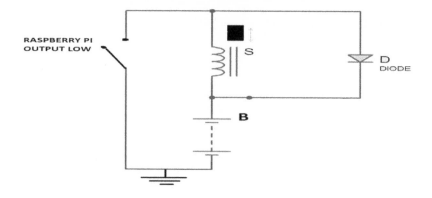

Fig no 5.2 Solenoid de-energized during OFF condition

At normal condition the solenoid actuator remains in the de-energized state. Because the solenoid actuator does not any signal from the control unit. Since the circuit is in the open condition there will be no power drawn from the power supply. So the solenoid actuator remains in the normal state.

5.2.2 CLOSED CIRCUIT CONDITION OF SOLENOID ACTUATOR

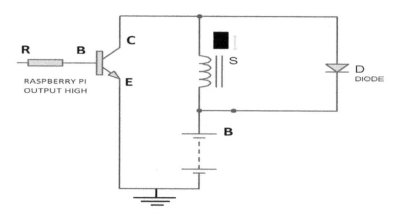

Fig no 5.3 Solenoid energized during ON condition

When the control signal is produced by the Raspberry pi, its output is fed to the base of the transistor. when the output of the Raspberry pi GPIO pins is fed to the base of the transistor,

31

it is triggered and the circuit is closed. When the circuit is closed solenoid actuator draws power from the power supply. Thus the solenoid actuator is energized. The same procedure is followed for all six solenoid actuators and thus it produces braille combinations for alphabets, numbers and punctuations in it.

5.3 CIRCUIT DIAGRAM OF POWER SUPPLY

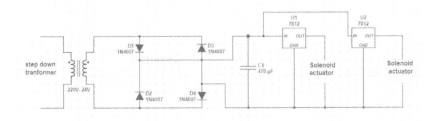

Fig no 5.4 circuit diagram of power supply

A rectifier is an electrical device that converts alternating current (AC), which periodically reverses direction, to direct current (DC), which flows in only one direction. The process is known as rectification.

5.3.1 IC 7812

Input voltage range: 15V – 35V

Current rating: 1.5A

Output voltage:12V

The difference between the input and output voltage appears as heat. The greater the difference between input and output voltage, the more heat is generated. If too much heat, is generated, the regulator gets over heated. If the regulator does not have a heat sink to dissipate this heat, it may damage. So, design the circuit such that the input voltage which is to be regulated is limited to 2-3 volts above the output regulated voltage or place an appropriate heat sink that can efficiently dissipate heat.

5.3.2 CAPACITOR

Capacitors are needed to filter the residual AC noise. Voltage regulators work efficiently work efficiently on a clean DC signal being fed. The bypass capacitors help to reduce the AC ripple. The heavier the load(higher current), capacitor discharges fastly.

$$C=\frac{I \times t}{\Delta v}$$

$$C=\frac{1 \times 0.01}{21}$$

C= 476uF

Since the calculated capacitance range 476uF, here in this circuit, a nearer value of 470uF is being used. When the load is so heavy that the ripple is too large. The bigger the value of the capacitor, the line is more smooth.

CHAPTER 6

TEXT TO VOICE SYTHESIS

6.1 INTRODUCTION

Talking computers were on fire 20 to 30 years ago. Movies like War Games and TV series like Knight Rider featured electronics speaking to their human operators. Speak-n-Spell machines taught a generation of children. Then the magic was out of the bottle and the focus drifted to other technologies. Today, we have a new explosion of voice enabled devices.

There are more devices speaking with the growth of the Internet of Things. Special speech synthesis chips of old are no longer required. The smallest of today's Linux computers often has the capabity to output speech. This includes the Raspberry Pi line of single board computers. From the Raspberry Pi Zero to the A+/B+ to the Raspberry Pi 2, all have the capacity to run free software to turn text to speech.

6.2 SOFTWARE PACKAGES

6.2.1 FESTIVAL

Festival, written by The Centre for Speech Technology Research in the UK, offers a framework for building speech synthesis systems. It offers full text to speech through a number APIs: from shell level, via a command interpreter, as a C++ library, from Java, and an Emacs editor interface. Festival is multi-lingual (currently British English, American English, and Spanish. Other groups work to release new languages for the system. Festival is in the package manager for the Raspberry Pi making it very easy to install.

6.2 2 FLITE

Flite is a lighter version of Festival built specifically for embedded systems. It has commands that make it easier to use than Festival on the command line. It runs faster than Festival. Unless you have the need for Festival's complex scripting language or phoneme handling, flite is the go-to program. Flite is also in the package manager. There are other speech programs for Linux, including espeak. If you find that Festival / Flite does not meet your needs, then check out espeak or other packages. Flite is designed as an alternative text to speech synthesis engine to Festival for voices built using the FestVox suite of voice building tools.

6.2.3 Espeak

Espeak is a compact open source software speech synthesizer for English and other languages, for Linux and Windows. Espeak uses a "formant synthesis" method. This allows many languages to be provided in a small size. The speech is clear, and can be used at high speeds, but is not as natural or smooth as larger synthesizers which are based on human speech recordings.

Espeak is available as:

- A command line program (Linux and Windows) to speak text from a file or from stdin.
- A shared library version for use by other programs. (On Windows this is a DLL).
- A SAPI5 version for Windows, so it can be used with screen-readers and other programs that support the Windows SAPI5 interface.
- espeak has been ported to other platforms, including Android, Mac OSX and Solaris.

6.3 AUDIO CONFIGURATION

The Raspberry Pi has two audio output modes: HDMI and headphone jack. You can switch between these modes at any time. If your HDMI monitor or TV has built-in speakers, the audio can be played over the HDMI cable, but you can switch it to a set of headphones or other speakers plugged into the headphone jack. If your display claims to have speakers, sound is output via HDMI by default; if not, it is output via the headphone jack. This may not be the desired output setup, or the auto-detection is inaccurate, in which case you can manually switch the output.

Changing the audio output

There are three ways of setting the audio output.

6.3.1 DESKTOP VOLUME CONTROL

Right-clicking the volume icon on the desktop taskbar brings up the audio output selector; this allows you to select between the internal audio outputs. It also allows you to select any external audio devices, such as USB sound cards and Bluetooth audio devices. A green tick is shown against the currently selected audio output device — simply left-click the desired output in the pop-up menu to change this. The volume control and mute operate on the currently selected device.

6.3.2 COMMAND LINE

The following command, entered in the command line, will switch the audio output to HDMI:

amixer cset numid=3 2

Here the output is being set to 2, which is HDMI. Setting the output to 1switches to analogue (headphone jack). The default setting is 0 which is automatic.

6.3.3 RASPI-CONFIG

Open up raspi-config by entering the following into the command line:

sudo raspi-config

This will open the configuration screen:

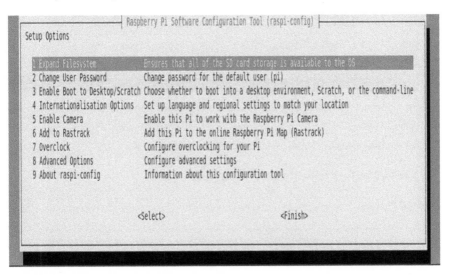

Select Option 7Advanced Options and press Enter.

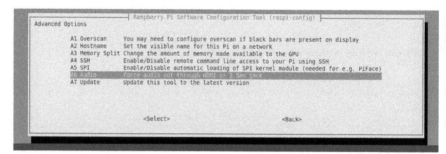

```
┌─────────── Raspberry Pi Software Configuration Tool (raspi-config) ┤
│
│  1 Change User Password       Change password for the default use
│  2 Hostname                   Set the visible name for this Pi on
│  3 Boot Options               Configure options for start-up
│  4 Localisation Options       Set up language and regional settin
│  5 Interfacing Options        Configure connections to peripheral
│  6 Overclock                  Configure overclocking for your Pi
│  7 Advanced Options           Configure advanced settings
│  8 Update                     Update this tool to the latest vers
│  9 About raspi-config         Information about this configuratio
│
│
│              <Select>                          <Finish>
│
└──────────────────────────────────────────────────────────────────
```

Then select Option A6: Audio and press Enter

```
┌──────────────── Raspberry Pi Software Configuration Tool (raspi-config) ┤
│ Advanced Options
│
│     A1 Overscan       You may need to configure overscan if black bars are present on display
│     A2 Hostname       Set the visible name for this Pi on a network
│     A3 Memory Split   Change the amount of memory made available to the GPU
│     A4 SSH            Enable/Disable remote command line access to your Pi using SSH
│     A5 SPI            Enable/Disable automatic loading of SPI kernel module (needed for e.g. PiFace)
│     A6 Audio          Force audio out through HDMI or 3.5mm jack
│     A7 Update         Update this tool to the latest version
│
│
│              <Select>                          <Back>
│
└──────────────────────────────────────────────────────────────────
```

Now you are presented with the two modes explained above as an alternative to the default Auto option. Select a mode, press Enter and press the right arrow key to exit the options list, then select Finish to exit the configuration tool.

6.4INSTALLATION FLITE PACKAGAE

Flite (festival-lite) is a small, fast open source text to speech synthesis engine developed at CMU and primarily designed for small embedded machines and/or large servers. Flite is designed as an alternative text to speech synthesis engine to Festival for voices built using the FestVox suite of voice building tools.

Flite is also included in the Debian package library so installation is similar too Festival. At the command prompt type:

sudo apt-get install flite

37

And you should see something like:

```
pi@raspberrypi:~ $ sudo apt-get install flite
Reading package lists... Done
Building dependency tree
Reading state information... Done
The following NEW packages will be installed:
  flite
0 upgraded, 1 newly installed, 0 to remove and 0 not upgraded.
Need to get 234 kB of archives.
After this operation, 384 kB of additional disk space will be used.
Get:1 http://mirrordirector.raspbian.org/raspbian/ jessie/main flite armhf 1.4-r
elease-12 [234 kB]
Fetched 234 kB in 4min 0s (973 B/s)
Selecting previously unselected package flite.
(Reading database ... 123821 files and directories currently installed.)
Preparing to unpack .../flite_1.4-release-12_armhf.deb ...
Unpacking flite (1.4-release-12) ...
Processing triggers for man-db (2.7.0.2-5) ...
Processing triggers for install-info (5.2.0.dfsg.1-6) ...
Setting up flite (1.4-release-12) ...
pi@raspberrypi:~ $
```

You are now set to try it out. For simple text, use text after the -t flag:

flite -t "All good men come to the aid of the rebellion"

and you can have flite speak the contents of a file with -f

flite -f jefferson-inaugural.txt

Sounds pretty good, eh?

You are not limited to the default voice. If you type

flite -lv

you get a list of available voices like this:

```
pi@raspberrypi:~ $ flite -lv
Voices available: kal awb_time kal16 awb rms slt
pi@raspberrypi:~ $
```

CHAPTER 7

TECHNICAL CHANLLENGES ENCOUNTERED DURING THE BRAILLE SYSTEM DESIGN

7.1 POWER SUPPLY DESIGN

Since the solenoid actuator draws more power it will not energize with the power supplied by the raspberry pi. The power requirement for the solenoid actuator was high and hence designing the external power supply for the solenoid actuator was difficult.

7.2 BRAILLE CHARACTER SIZE

Distance between dots

- Horizontal or vertical distance from centre to centre of adjacent dots in the same cell = 2.3 mm – 2.5 mm.
- Horizontal distance from centre to centre of corresponding dots in adjacent cells = 6.0 mm – 7.0 mm.
- Vertical line spacing from centre to centre of nearest corresponding dots in adjacent lines = 10.0 mm – 11.0 mm.

Dot size

- dot height = 0.6 mm – 0.9 mm.
- dot base diameter = 1.5 mm – 1.6 mm.
- spherical radius = 0.76 mm – 0.81 mm.

These are the braille character size but here the braille display is solenoid actuator which is much bigger than the actual braille size. This may affect the users because they learn the braille combinations in bigger size and actual braille books have braille combinations in smaller size. To overcome this problem, an arrangement was made. Hence to reduce the size of the braille combinations to the actual size what blind people are using, we are here attaching a small thin iron rod with the solenoid actuator.

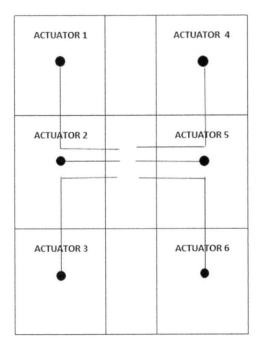

Fig no 7.1 Arrangement of braille cell for the smaller size

7.3 ADVANTAGES OF THE BRAILLE KIT

➢ Self-learning kit

➢ Low cost

➢ Can learn braille faster

➢ Can be used in schools to teach braille for blind and visually impaired people

CONCLUSION

Braille teaching kit is low in cost and is affordable for all class of people. This kit will help the blind and visually impaired to learn braille by their own in an efficient manner. This kit can be used numerous number of times and hence the user can learn braille in their own phase, they need not to worry about the time limit. This can be used in braille teaching schools which will be best replacement for lack of good skilled teachers.

FUTURE SCOPE

The product can be enhanced with addition of extra features like automatic search mode where the blind people can search a letter by giving the voice input, that is they can give input of any of the alphabets and numbers and the corresponding braille output is produced and in the testing mode the braille combination will be produced in the braille kit, the user had to touch and feel the braille combinations and say the which alphabet or number the combination belong to, this will enhance the remembrance level of the blind and visually impaired people.

HARDWARE MODULE

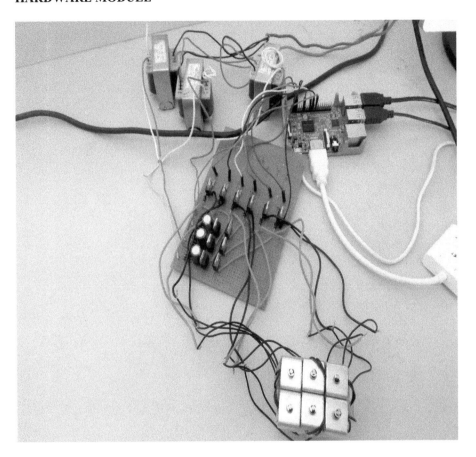

APPENDIX

```
import time
import RPi.GPIO as GPIO
GPIO.cleanup()
GPIO.setmode(GPIO.BOARD)
GPIO.setup(3,GPIO.OUT)
GPIO.setup(5,GPIO.OUT)
GPIO.setup(13,GPIO.OUT)
GPIO.setup(19,GPIO.OUT)
GPIO.setup(26,GPIO.OUT)
GPIO.setup(21,GPIO.OUT)
while(1):
    print('letter a')
    GPIO.output(3,True)
    GPIO.output(5,False)
    GPIO.output(13,False)
    GPIO.output(19,False)
    GPIO.output(26,False)
    GPIO.output(21,False)
    time.sleep(3)
    GPIO.output(3,False)
    time.sleep(2)
    print('letter b')
    GPIO.output(3,True)
    GPIO.output(5,True)
    GPIO.output(13,False)
    GPIO.output(19,False)
    GPIO.output(26,False)
    GPIO.output(21,False)
    time.sleep(3)
    GPIO.output(3,False)
    GPIO.output(5,False)
    time.sleep(2)
```

```python
print('letter c')
GPIO.output(3,True)
GPIO.output(5,False)
GPIO.output(13,False)
GPIO.output(19,True)
GPIO.output(26,False)
GPIO.output(21,False)
time.sleep(3)
GPIO.output(3,False)
GPIO.output(19,False)
time.sleep(2)
print('letter d')
GPIO.output(3,True)
GPIO.output(5,False)
GPIO.output(13,False)
GPIO.output(19,True)
GPIO.output(26,True)
GPIO.output(21,False)
time.sleep(3)
GPIO.output(3,False)
GPIO.output(19,False)
GPIO.output(26,False)
time.sleep(2)
print('letter e')
GPIO.output(3,True)
GPIO.output(5,False)
GPIO.output(13,False)
GPIO.output(19,False)
GPIO.output(26,True)
GPIO.output(21,False)
time.sleep(3)
GPIO.output(3,False)
GPIO.output(26,False)
time.sleep(2)
```

```
print('letter f')
GPIO.output(3,True)
GPIO.output(5,True)
GPIO.output(13,False)
GPIO.output(19,True)
GPIO.output(26,False)
GPIO.output(21,False)
time.sleep(3)
GPIO.output(3,False)
GPIO.output(5,False)
GPIO.output(19,False)
time.sleep(2)
print('letter g')
GPIO.output(3,True)
GPIO.output(5,True)
GPIO.output(13,False)
GPIO.output(19,True)
GPIO.output(26,True)
GPIO.output(21,False)
time.sleep(3)
GPIO.output(3,False)
GPIO.output(19,False)
GPIO.output(5,False)
GPIO.output(26,False)
time.sleep(2)
print('letter h')
GPIO.output(3,True)
GPIO.output(5,True)
GPIO.output(13,False)
GPIO.output(19,False)
GPIO.output(26,True)
GPIO.output(21,False)
time.sleep(3)
GPIO.output(3,False)
```

```
GPIO.output(5,False)
GPIO.output(26,False)
time.sleep(2)
print('letter i')
GPIO.output(3,False)
GPIO.output(5,True)
GPIO.output(13,False)
GPIO.output(19,True)
GPIO.output(26,False)
GPIO.output(21,False)
time.sleep(3)
GPIO.output(5,False)
GPIO.output(19,False)
time.sleep(2)
print('letter j')
GPIO.output(3,False)
GPIO.output(5,True)
GPIO.output(13,False)
GPIO.output(19,True)
GPIO.output(26,True)
GPIO.output(21,False)
time.sleep(3)
GPIO.output(19,False)
GPIO.output(26,False)
GPIO.output(5,False)
time.sleep(2)
print('letter k')
GPIO.output(3,True)
GPIO.output(5,False)
GPIO.output(13,True)
GPIO.output(19,False)
GPIO.output(26,False)
GPIO.output(21,False)
time.sleep(3)
```

```python
GPIO.output(3,False)
GPIO.output(13,False)
time.sleep(2)
print('letter l')
GPIO.output(3,True)
GPIO.output(5,True)
GPIO.output(13,True)
GPIO.output(19,False)
GPIO.output(26,False)
GPIO.output(21,False)
time.sleep(3)
GPIO.output(3,False)
GPIO.output(5,False)
GPIO.output(13,False)
time.sleep(2)
print('letter m')
GPIO.output(3,True)
GPIO.output(5,False)
GPIO.output(13,True)
GPIO.output(19,True)
GPIO.output(26,False)
GPIO.output(21,False)
time.sleep(3)
GPIO.output(3,False)
GPIO.output(13,False)
GPIO.output(19,False)
time.sleep(2)
print('letter n')
GPIO.output(3,True)
GPIO.output(5,False)
GPIO.output(13,True)
GPIO.output(19,True)
GPIO.output(26,True)
GPIO.output(21,False)
```

```
time.sleep(3)
GPIO.output(3,False)
GPIO.output(13,False)
GPIO.output(19,False)
GPIO.output(26,False)
time.sleep(2)
print('letter o')
GPIO.output(3,True)
GPIO.output(5,False)
GPIO.output(13,True)
GPIO.output(19,False)
GPIO.output(26,True)
GPIO.output(21,False)
time.sleep(3)
GPIO.output(3,False)
GPIO.output(13,False)
GPIO.output(26,False)
time.sleep(2)
print('letter p')
GPIO.output(3,True)
GPIO.output(5,True)
GPIO.output(13,True)
GPIO.output(19,True)
GPIO.output(26,False)
GPIO.output(21,False)
time.sleep(3)
GPIO.output(3,False)
GPIO.output(13,False)
GPIO.output(5,False)
GPIO.output(19,False)
time.sleep(2)
print('letter q')
GPIO.output(3,True)
GPIO.output(5,True)
```

```
GPIO.output(13,True)
GPIO.output(19,True)
GPIO.output(26,True)
GPIO.output(21,False)
time.sleep(3)
GPIO.output(3,False)
GPIO.output(13,False)
GPIO.output(5,False)
GPIO.output(19,False)
GPIO.output(26,False)
time.sleep(2)
print('letter r')
GPIO.output(3,True)
GPIO.output(5,True)
GPIO.output(13,True)
GPIO.output(19,False)
GPIO.output(26,True)
GPIO.output(21,False)
time.sleep(3)
GPIO.output(3,False)
GPIO.output(13,False)
GPIO.output(5,False)
GPIO.output(26,False)
time.sleep(2)
print('letter s')
GPIO.output(3,False)
GPIO.output(5,True)
GPIO.output(13,True)
GPIO.output(19,True)
GPIO.output(26,False)
GPIO.output(21,False)
time.sleep(3)
GPIO.output(5,False)
GPIO.output(13,False)
```

```
GPIO.output(19,False)
time.sleep(2)
print('letter t')
GPIO.output(3,False)
GPIO.output(5,True)
GPIO.output(13,True)
GPIO.output(19,True)
GPIO.output(26,True)
GPIO.output(21,False)
time.sleep(3)
GPIO.output(13,False)
GPIO.output(5,False)
GPIO.output(19,False)
GPIO.output(26,False)
time.sleep(2)
print('letter u')
GPIO.output(3,True)
GPIO.output(5,False)
GPIO.output(13,True)
GPIO.output(19,False)
GPIO.output(26,False)
GPIO.output(21,True)
time.sleep(3)
GPIO.output(3,False)
GPIO.output(13,False)
GPIO.output(21,False)
time.sleep(2)
print('letter v')
GPIO.output(3,True)
GPIO.output(5,True)
GPIO.output(13,True)
GPIO.output(19,False)
GPIO.output(26,False)
GPIO.output(21,True)
```

```
time.sleep(3)
GPIO.output(3,False)
GPIO.output(13,False)
GPIO.output(5,False)
GPIO.output(21,False)
time.sleep(2)
print('letter w')
GPIO.output(3,False)
GPIO.output(5,True)
GPIO.output(13,False)
GPIO.output(19,True)
GPIO.output(26,True)
GPIO.output(21,True)
time.sleep(3)
GPIO.output(5,False)
GPIO.output(13,False)
GPIO.output(19,False)
GPIO.output(26,False)
GPIO.output(21,False)
time.sleep(2)
print('letter x')
GPIO.output(3,True)
GPIO.output(5,False)
GPIO.output(13,True)
GPIO.output(19,True)
GPIO.output(26,False)
GPIO.output(21,True)
time.sleep(3)
GPIO.output(3,False)
GPIO.output(13,False)
GPIO.output(19,False)
GPIO.output(21,False)
time.sleep(2)
print('letter y')
```

```
GPIO.output(3,True)
GPIO.output(5,False)
GPIO.output(13,True)
GPIO.output(19,True)
GPIO.output(26,True)
GPIO.output(21,True)
time.sleep(3)
GPIO.output(3,False)
GPIO.output(13,False)
GPIO.output(19,False)
GPIO.output(26,False)
GPIO.output(21,False)
time.sleep(2)
print('letter z')
GPIO.output(3,True)
GPIO.output(5,False)
GPIO.output(13,True)
GPIO.output(19,False)
GPIO.output(26,True)
GPIO.output(21,True)
time.sleep(3)
GPIO.output(3,False)
GPIO.output(13,False)
GPIO.output(26,False)
GPIO.output(21,False)
time.sleep(2)
```

REFERENCE

1. Saurabh Bisht, Sandeep Reddy Goluguri , Rajat Maheshwari, Akhilesh Kumar and P.Sathya "Refreshable Braille Display using Raspberry Pi and Arduino", International Journal of Current Engineering and Technology,Accepted 01 June 2016, Available online 06 June 2016, Vol.6, No.3 (June 2016)

2. Prachi Rajarapollu, Stavan Kodolikar, Dhananjay Laghate, Amarsinh Khavale , "FPGA Based Braille to Text & Speech For Blind Persons", International Journal of Scientific & Engineering Research, Volume 4, Issue 4, April-2013 348 ISSN 2229-5518

3. Aisha Mousa, HazeemHiary, Alomari, and Loai Alnemer "Firstland Experience with balanced Approach"2009 International Conference on, no., pp.201-211,17-19 April 2009.

4. Vineeth Kartha; Dheeraj S. Nair; S. Sreekant; P. Pranoy; P. Jayaprakash(2012) "DRISHTI – A gesture controlled text to braille converter", India Conference(INDICON),2012 Annual IEEE, pp. 335-339.

5. TechBridgeWorld Automated Braille Writing Tutor project website: http://www.techbridgeworld.org/brailletutor.

6. World Health Organization, "Fact sheet 282: Visual impairment and blindness," World Health Organization, May 2009

7.https://en.wikipedia.org/wiki/Braille

YOUR KNOWLEDGE HAS VALUE

- We will publish your bachelor's and master's thesis, essays and papers

- Your own eBook and book - sold worldwide in all relevant shops

- Earn money with each sale

Upload your text at www.GRIN.com and publish for free